CW00557573

TROPICAL WILDLIFE OF SOUTHEAST ASIA

PERIPLUS NATURE GUIDES

TROPICAL WILDLIFE

of Southeast Asia

Text by Jane Whitten

Photography by Alain Compost

PERIPLUS

Published by Periplus Editions (HK) Ltd.

Printed in Singapore
ISBN 962-593-065-5

Publisher: Eric M. Oey
Design: Peter Ivey
Cover photography: Alain Compost
Additional photography: Gerald Cubitt, p. 17 (top), p. 32 (top) and p. 33 (top); Tim Flannery, p. 39 (top); Francis Lim, p. 52 (bottom), p. 53 (bottom), p. 56 (top) and all photos on pp. 58, 59; and Kal Muller, p. 54 (top) and p. 55 (top).
Editor: Julia Henderson
Production: TC Su

Distributors

Indonesia
PT Wira Mandala Pustaka,
(Java Books - Indonesia),
Jalan Kelapa Gading Kirana,
Blok A14 No. 17,
Jakarta 14240

Singapore and Malaysia
Berkeley Books Pte. Ltd.,
5 Little Road #08-01, Singapore 536983

North America, Latin America and Europe
Tuttle Publishing
Distribution Center
Airport Industrial Park
364 Innovation Drive
North Clarendon, VT 05759-9436

Introduction

The Southeast Asian countries of Thailand, Malaysia and Indonesia lie within one of the most biologically rich regions on Earth. Glaciers, extensive swamps, savanna, grassy plains, tall mountains and many different types of forest cover this remarkable area. Tropical forests are the dominant natural vegetation and each forest type has its own characteristic species. As a group, these forests are known to harbor one of the most diverse assemblages of wildlife in the world.

Three well-defined zoogeographical realms lie within Southeast Asia: the Oriental region (which extends from Pakistan across southern China, Taiwan, the Philippines and into Indonesia); the Australasian region (reaching from the island of New Guinea, south-east across Australia including New Zealand); and a transitional subregion, known as "Wallacea," located between the two realms (covering central Indonesia). The animals found in the Oriental region are strikingly different from those in the Australasian region, by virtue of the very different geological histories of these regions.

An incredible variety of wildlife occurs in Southeast Asia—altogether about 550 species of mammals, over 500 species of reptiles and over 500 species of amphibians. Some animals, such as bats and geckos, are widespread throughout the region and are easy to find. Other widespread species such as jungle cats, otters and blind snakes can be sighted only with luck and diligence. A few species may be observed only in a single national park which encompasses almost their entire range. Of these, the Komodo dragon is very easy to watch, while it is extremely difficult to catch even a fleeting glimpse of the Javan rhinoceros. Other species with relatively small ranges, such as the anoa, Bawean deer and false ghavial are perhaps best seen in zoos.

The remote areas of Southeast Asia are becoming increasingly accessible as more national parks are developed, and road and air links to them are improved. These national parks, many with excellent interpretive centers, are fostering an increasing awareness and appreciation of the region's unique wildlife. Many species are in danger of extinction through the relentless destruction of forest habitats by logging and land clearance for agricultural and plantation purposes. Land clearance is often facilitated by widespread, damaging fires.

This Periplus Nature Guide covers mammals, reptiles and amphibians and offers an excellent introduction to a wide variety of Southeast Asia's common, well-known species, as well as some of the more extraordinary animals that have evolved in restricted habitats. The region's birdlife, reef fishes and invertebrate marine species are covered in separate volumes in this series.

—Jane Whitten

Orangutans

Above and opposite:
Orangutan,
Pongo pygmaeus

These large, red, long-haired apes are found in the forests of lowland North Sumatra and Borneo. The impressive adult male, with cheek flanges of fibrous tissue accentuating his face, can weigh up to 90 kg. Females are smaller, weighing only 40–50 kg.

These diurnal apes spend most of their time in the trees. They live on fruit, some of which is poisonous to humans, supplemented by leaves and shoots, but will also eat eggs and small animals. Their detailed knowledge of the forest enables them to search very efficiently for fruiting trees. Orangutans travel by holding on with their hook-like hands and feet, using their weight to swing trees back and forth to narrow a gap so that they can reach the next tree. Travel is cumbersome, and they go only a few hundred meters each day, fashioning branches into a treetop nest at dusk. Older, heavier males may resort to traveling on the ground, giving rise to stories of wild forest men.

Orangutans are largely solitary, although young animals sometimes play together. Females mature at ten years and may have three or four offspring during their lifetimes, weaning them at about three years. Juveniles finally leave their mothers at 7–10 years of age. Adult males defend those females within their range from other males, advertising their presence by calling, displaying aggressively when they meet and occasionally even fighting.

These animals are extremely strong and able to crack hard forest fruits easily, but are also dexterous and very curious. A camera inadvertently left behind, which caught the attention of wild orangutans, was later found totally dismantled. Nothing was broken.

Orangutans are threatened by habitat loss and hunting as females are sometimes killed to obtain infants for the pet trade. However, large reserves in Indonesia and Malaysia contain good populations of orangutans, and a prolonged campaign of public education has lessened the pet trade.

Above:
Mentawai gibbon,
Hylobates klossii

Opposite

Top left and bottom right:
Javan gibbon,
Hylobates moloch

Top right:
Siamang,
Hylobates syndactylus

Center and bottom left:
Müller's gibbon,
Hylobates muelleri

Gibbons

There are nine species of gibbon in the forests of Southeast Asia. Eight of these species weigh around 6 kg, and the larger siamang weighs about 10 kg. The smaller species are geographically separated by seas and rivers, but the siamangs overlap with them in the forests of Peninsular Malaysia and North Sumatra. Gibbons are uniquely adapted to life in the forest canopy and are almost never found on the ground. They travel quickly and efficiently by swinging from branch to branch, often leaping between supports. The siamang's diet consists almost equally of fruit and leaves, while the smaller gibbons eat mainly fruit, which they pick carefully, leaving unripe fruit to ripen. They also eat leaves, shoots and insects.

Gibbons live in monogamous families into which a single offspring is born every two or three years. Infants are weaned in their second year, but stay with their parents until they are about eight. In siamang groups, the male takes care of the infant after the first year.

A gibbon family lives within a well-defined home territory, which the animals know well, visiting trees as they come into fruit. A gibbon group travels about 1.5 km in a day whereas the larger siamangs travel less than a kilometer. The groups sleep in tall, emergent trees, safe from predators. Before dawn, the males of some species sing to advertise the presence of the group and their occupancy of that area of forest. About a third of the active day is spent feeding and another third traveling between feeding trees.

Siamang groups are very cohesive, with individuals rarely more than 10 m from each other, whereas individuals in a group of small gibbons are more spread out, coming together to feed in larger fruiting trees. Adults groom each other, which helps maintain the pair bond. Most gibbon pairs also perform an elaborate duet that lasts for about 15 minutes. These stereotyped songs differ between species and are thought to advertise the presence of a pair in the territory and to reinforce pair bonding.

Above:
Silvered leaf monkey,
Semnopithecus cristata

Opposite

Top left:
Female and juvenile proboscis monkey,
Nasalis larvatus

Bottom left:
Male proboscis monkey,
Nasalis larvatus

Top and bottom right:
Banded leaf monkey,
Presbytis melalophos

Center right:
Javan leaf monkey,
Presbytis comata

Proboscis and other Leaf Monkeys

Leaf monkeys are found through most of Southeast Asia although some species, like the proboscis monkey, have very restricted ranges. Most are relatively slender monkeys with long tails and largely arboreal lifestyles. The species are distinguished by coat color and hair patterns on the head.

Some leaf monkeys eat nothing but leaves, but most eat flowers, buds, seeds and shoots as well. They have complex stomachs, with bacteria-filled fermenting chambers that break down leaves, releasing the normally indigestible sugars and deactivating the leaf toxins. Because the leaf monkey eats such poor quality food, it has to forage for much of the day, and its stomach contents may constitute a quarter of the animal's weight.

Leaf monkeys live mostly in large groups comprising one male and a number of females with their offspring. An infant is born with its eyes open and is strong enough to hold onto the mother's fur while she travels. Females frequently tolerate and even suckle each other's small infants. Excess males form groups and look for opportunities to take over from breeding males. Incoming males usually kill all the infants fathered by their predecessors, although the mothers sometimes manage to protect their infants.

The proboscis monkey is found only in the coastal swamp forests of Borneo. The males have remarkable faces, with long pendulous noses, the functions of which have long been the source of speculation. In the past, suggestions have been made that they enable the large male to give off heat or that the noses assist in swimming by acting like some kind of snorkel. It is now thought that they serve to attract females, like the peacock's tail. Proboscis monkeys feed mostly on young leaves and travel further than most forest monkeys to obtain this scarce food. These monkeys swim well and cross large rivers very quietly, with no splashing that could attract crocodiles.

Above:
Booted macaque,
Macaca ochreata

Opposite

Top left:
Long-tailed macaque,
Macaca fascicularis

Top and center right:
Black-crested macaque,
Macaca nigra

Bottom:
Pig-tailed macaque,
Macaca nemestrina

Macaques

Macaques are the typical monkey—gregarious, active an curious. They are primarily fruit eaters, but will also ea other plant parts and small animals, from insects to bir and mammals, if they can catch them. Where they com into contact with people, they will expand their diets t include garbage, peanuts and other offerings, or stole crops. They spend more time on the ground than the lea monkeys, but are excellent climbers and swimmers.

Macaques are long-lived and give birth to a singl infant each year, after a gestation of five or six months The infant is born with fur and open eyes, and can imme diately travel clinging to its mother's belly fur, with a nip ple in its mouth to support its head. Nursing become infrequent after the first few months, but usually contin ues until the next infant is born. These social animal spend long periods in mutual grooming, which helps rein force group cohesion.

The long-tailed or crab-eating macaque is found in variety of habitats, especially coastal forests and offshor islands where it frequents low trees and scrub. The pig tailed macaque, named for the shorter tail which it carrie curled over its back, is found more in inland hill forests. I is mostly arboreal, but is unique in that it will descend t the ground to flee from people. Pig-tailed macaques seen to learn more readily than other macaques, and captiv pig-tails are trained throughout the region to pic coconuts. Experienced individuals are able to choose rip fruit without help, while less experienced individuals ar directed by their handlers. Both species are found in groups of up to forty individuals, but these large group often split up to feed.

On the island of Sulawesi there are four distinc species of macaque, all of which are found nowhere else They all resemble the pig-tailed macaque, but differ in fur pattern and color.

Above:
Slow loris,
Nycticebus coucang

Opposite

Top left: Slow loris,
Nycticebus coucang

Top right:
Western tarsier,
Tarsius bancanus

Bottom:
Spectral tarsier,
Tarsius spectrum

Lorises and Tarsiers

The slow loris is very appropriately named. This smal nocturnal creature prowls around the lower and middl canopy of mature or disturbed forest in search of frui nestlings and other small animals. It climbs very delib erately and slowly, walking hand over hand along th branches, one limb at a time and is able to remai motionless for hours on end. When catching an insec the slow loris stands on its hind feet and throws it body forward with surprising speed, grabbing the ani mal with both hands and dispatching it with a bit from its sharp teeth.

Female lorises do not make nests, but give birth in th open to a single young each year. The offspring is at firs carried continuously, but is later parked on a branch fo short periods while the mother hunts. The parents usuall sleep close together and the young may be found clingin to either of them.

Tarsiers can be found in all sorts of habitats includin towns, mangrove forests, secondary forests and montan forests. These tiny primates, weighing no more than 135 g have enormous eyes. Each eye weighs a little more tha the whole brain, giving them excellent night vision fo hunting. They catch beetles, cockroaches, ants, birds an sometimes snakes by leaping at them and pinning then down with one or both hands, often on the ground.

Tarsiers are most active in the hour or so befor dawn. The adults and their offspring head back to thei sleeping sites in thickets among vine or fig tangles, or i tree holes. Just before retiring the family sings a comple call notifying all those with an interest in the matter tha they are still alive and defending their one-hectare terri tory, the boundaries of which they will have marked ou with drops of odoriferous urine during the night.

Wild Cattle

Above

Top:
Lowland anoa,
Bubalus depressicornis

Bottom:
Mountain anoa,
Bubalus quarlesi

Opposite

Top:
Gaur herd,
Bos gaurus

Bottom:
Banteng herd,
Bos javanicus

The two species of diminutive, dark buffalo endemic to Sulawesi are called anoas. They eat young leaves, fruit, grasses and ferns. This diet is very low in salt so they use mineral licks and sometimes drink sea water. Anoas need undisturbed forest and cannot adapt to logged areas. The bad-tempered anoa with its sharp short horns is greatly feared, but in spite of this it is periodically suggested as a potential domestic animal. Anyone who has met with an anoa ridicules this idea.

Banteng are large, handsome cattle, weighing up to 800 kg. The male is a dark chestnut brown and the female mid-brown. Both have a contrasting white rump patch and stockings. They are the ancestors of the domestic Bali cattle, and young wild animals look almost identical to the domestic variety. Banteng graze on grasses, preferring open areas such as clearings and river banks to dense forest. The cows and calves generally live in herds of twenty-five or more, with one adult male. Each group is led by an older cow. Surplus males form bachelor herds. Herding helps protect these animals from leopards and tigers. Wild banteng are largely nocturnal, possibly to avoid large diurnal predators. They can adapt to living in logged forest, as many of their food plants grow on open ground, but cannot survive forest clearance. Banteng also face the threat of interbreeding with domestic Bali cattle and hence losing their genetic identity.

Gaur are massive beasts, weighing up to 900 kg, with white stockings on their legs and inward curving horns like water buffalo. Calves are born brown and change color at four or five months. Gaur live in herds of up to twenty and are often found in deep forest where they browse on the leaves of herbs and saplings. They are mostly nocturnal, lying in deep shade during the heat of the day. A wounded gaur bull is considered very dangerous and will charge and toss its attacker.

Above:
Sambar deer,
Cervus unicolor

Opposite

Top left:
Lesser mouse deer,
Tragulus javanicus

Top right and bottom:
Rusa deer,
Cervus timorensis

Center left:
Muntjac deer,
Muntiacus muntjac

Deer

The smallest of all the deer is the lesser mouse dee which weighs only 2 kg and stands 20 cm tall. This deer i thought to be the most primitive of the deer family—a intermediate between deer and pigs—since it has a shor neck, no antlers and protruding upper tusks in the mal This solitary deer is largely nocturnal and lives in dens rainforest, feeding on shoots, leaves and fallen fruit. Th greater and lesser mouse deer are both widesprea throughout the region—as far south as Sumatra and Jav respectively. The greater mouse deer weighs up to thre times as much as the lesser.

Male muntjac deer have short antlers and both male and females are fairly small, standing about 50 cm at th shoulder. They occur in all types of forest, and walk an run in a very distinctive way—with their heads held lo and their rumps high, lifting each foot high as they ste When alarmed they give a short series of loud bark hence their common name of barking deer.

The large, brown sambar deer are widesprea throughout the west of the region and are found in a var ety of habitats, from open woodland to dense forest. The are tolerant of man and are often seen close to village The sambar deer are quite similar to the rusa deer, whic occur in Java and eastern Indonesia. The two species di fer in the details of their antlers and in the rusa's prefe ence for more open habitats such as grassland. Samb deer browse, feeding on young leaves, buds, fallen fru and some grass, whereas rusa deer are grazers, feedin largely on grasses.

The Bawean deer is one of the rarest deer in th world, now found only on the small volcanic island Bawean, north of Java. This animal was once almo extinct, but protection has boosted numbers to 200–40 animals. However, uncontrolled fires and the conversio of the remaining steep forest land to dryland agricultu remain serious threats to its existence.

Above:
Young wild boar,
Sus scrofa

Opposite

Top:
Babirusa,
Babyrousa babyrussa

Bottom:
Wild boar,
Sus scrofa

Wild Pigs

The babirusa, found only in Sulawesi, seems to hav diverged from pig ancestry 30 million years ago. The pi usually digs in the soil for roots and worms, whereas th babirusa eats fruit found on the ground and breaks ope fallen tree trunks to obtain beetle larvae. Babirusa produce only one or two offspring at a time which grov relatively slowly, unlike pigs which have large litters o fast-growing young. The male's upper canine teet curve upward, then around toward the skull. It seem probable that these unusual tusks have a role in mal aggression. Babirusa are rare and becoming rarer despit full legal protection.

The warty pig, found only on Java, is similar to th common wild boar that occurs throughout the temperate and tropical regions. This pig, distinguished by th three pairs of warts on its face, is not as rare as was onc feared, and is now common in some areas. The warty pi prefers areas of lowland secondary vegetation, particularly teak plantations in Central Java consisting of a mixture of different-aged trees and grasslands with clump of bush or heavily disturbed forest. It also lives in coasta forests where it is often the only pig species present The warty pig roams in small groups of four to six whereas the wild boar can sometimes be seen in aggregations ten times this size.

The bearded pig sometimes forms large herds whic undertake mass migrations in response to the synchronous fruiting of dipterocarp trees. At these times the form a major part of the diet and economy of forest dwelling peoples in Borneo.

These pigs are all heavily hunted for pleasure and ar often snared and illegally poisoned to protect crops The meat is sold in local markets. The wild boar an bearded pig are still quite common in some areas, however, the babirusa is gravely threatened.

Tapirs and Rhinos

Above:
Juvenile Malay tapir,
Tapirus indicus

Opposite

Top left:
Javan rhinoceros,
Rhinoceros sondaicus

Top right:
Adult Malay tapir,
Tapirus indicus

Center right and bottom:
Sumatran rhinoceros,
Dicerorhinus sumatrensis

The tapir is found primarily in low-lying or swampy forests in Thailand, Malaysia and southern Sumatra. These striking animals are surprisingly well camouflaged in the thick forests they frequent. When threatened they move quickly and will hide in deep streams, staying submerged for long periods. Young tapirs are striped and the adult coloration develops after about four months.

A hundred years ago the Javan rhinoceros was found throughout much of Southeast Asia, but is now restricted to a few isolated populations. The best known is in Ujung Kulon National Park, in Java. This rhino stands 1.5 m at the shoulder and weighs up to two tons. It looks much like the Indian rhino, but has an extra fold of skin around the neck. All males have a short horn, averaging 15 cm but most females have only a suspicion of a bump.

The Sumatran rhinoceros used to be found throughout much of Southeast Asia, as far south as Sumatra. A few small populations survive in Malaysia and Vietnam but probably the only viable population remaining is in Sumatra. This rhino is much smaller than the Javan, and is considered the most primitive rhino, being quite hairy.

Javan rhinos are browsing animals which use thei long, prehensile upper lips to gather a wide range o young leaves, shoots, twigs and fallen fruit along the forest margins. This use of disturbed areas has brought them into contact with people, and hunting has driven these animals to extinction in almost all areas. The Sumatran rhino, in contrast, is found mostly in deep forest, where i feeds on woody plants.

Rhinos live largely solitary lives, forming adult pairs only temporarily at breeding time, and cow-calf pairs for the three or four years that the mother looks afte her young.

Above and opposite: Asian elephants, *Elephas maximus*

Asian Elephants

The Asian elephant can be distinguished from the African elephant by its smaller size, high back, shorter tusks (not visible outside the mouth in females and some males) and smaller ears. These elephants frequent lowland forests from India to Sumatra, and live in herds or extended families of five to twenty animals, led by an old female. Elephants give birth to one calf after about twenty months' gestation. During the birth the mother is attended by another female who will also help look after the youngster. Adult males join the herd only to mate.

Asian elephants feed on the fast-growing, succulent plants of lowland forest gaps and edges and are also partial to fruit, even knocking down trees to reach it. Elephants are sometimes accused of being wasteful, since they leave a trail of devastation as they feed. In the forest this broken vegetation becomes available to wild cattle and deer, which may follow behind an elephant herd for this very reason. Clearly elephants must be able to range over a large area so that damaged patches have time to recover. An adult elephant eats about 150 kg of food and drinks 70–90 l of water a day. Asian elephants are not tolerant of heat and will seek out deep shade in the middle of the day. Their small ears are less able to conduct heat away from their large bodies than are the larger ears of the African elephants.

Their liking for succulent vegetation and fruit brings conflict wherever people and wild elephants meet. In Indonesia the elephant is in the unusual situation of being classed both as a protected animal and as a pest. Many different methods have been used to protect farmers—from relocating the elephants to setting up electric fences—all with mixed success.

The domestication of elephants is an ancient tradition in Thailand, where their great strength, longevity and ability to learn make them valuable beasts of burden. However they are not a truly domestic species because they are rarely bred in captivity.

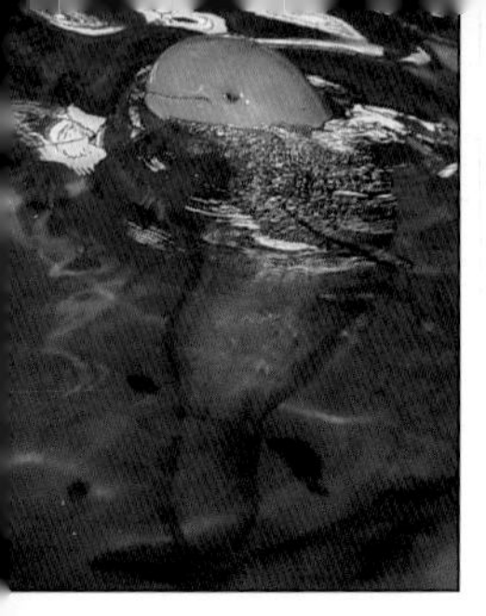

Above and opposite top: Irrawaddy dolphin, *Orcaella brevirostris*

Bottom: Dugong, *Dugong dugon*

Dolphins and Dugongs

Many species of dolphin are ocean-going, but some prefer shallow inshore waters and estuaries. The latter group includes the black finless porpoise (which is actually gray, but does indeed have no dorsal fin) and the Irrawaddy dolphin. The latter is a small, white dolphin with a rounded head and tiny eyes, which are adaptations to living in heavily silted water with poor visibility. It is thought to frequent the lower reaches of large rivers as well as coastal waters. Many dolphins feed on fish, but this species feeds on crayfish and shrimp, picked from the sea floor or riverbeds with its downward-pointing mouth.

The dugong, the only herbivorous marine mammal, used to be found in coastal waters from east Africa to the south Pacific. Dugongs are superficially similar to seals, but are more closely related to elephants. Like elephants, they have relatively inefficient digestive systems and must eat very large quantities of plant material to extract enough nutrients. They are threatened with extinction because of their slow reproductive rates, docile natures, valuable small tusks and delicious flesh.

Adult dugongs are 2–3 m long and weigh up to 400 kg. These lethargic creatures feed in the shallow coastal seagrass meadows. The carbohydrate-rich rhizomes found just below the surface of the sand make up the bulk of their food and they eat some seagrass leaves as well. Their lives are relaxed and easy since no other creatures feed on the same food—their closest competition comes from the green turtle which eats just the leaf blades of the seagrass. They appear to have no natural enemies except man, against whom they have no defense.

Female dugongs mature at eight to eighteen years of age and have a single young which is suckled for about two years. A female may live for fifty years, but is unlikely to have more than five or six young during her lifetime.

Above:
Clouded leopard,
Neofelis nebulosa

Opposite

Top left and bottom:
Sumatran tiger,
Panthera tigris

Top right and center:
Spotted leopard and black panther,
Panthera pardus

Tigers and Leopards

Three large cats are found in Southeast Asia—the spotted and clouded leopards and the tiger. Tigers and clouded leopards prefer a forest habitat, whereas spotted leopards are tolerant of less cover, even venturing into suburban fringes. All three have suffered from hunting for their skins and persecution for their feeding habits, but leopard populations are still healthy in some areas.

Tigers, the largest cats, inspire great fear and respect. Their long hind limbs, heavily muscled shoulders, retractable claws and large canine teeth are superbly adapted for catching and killing large prey. They may cover 10–20 km in a night in their search for food, hunting alone and capturing their prey only once in several tries. Three or four cubs are born in a den and raised by the female. They are dependent on her for food for eighteen months and remain with her until they are two and a half years old. Most tigers are afraid of people and avoid them. However, old or injured individuals may discover that people are easy prey and become man-eaters. These tigers are killed to protect the local human population.

Spotted leopards are very adaptable. They hunt at night and rest in trees during the day, capturing a variety of small prey such as birds and small mammals by a combination of speed and stealth. This leopard occurs in two distinct color phases, the spotted and the black. The latter is known as the black panther. On the Malay Peninsula as many as half the leopards are black. Litters usually consist of three cubs; these stay with the mother until they are eighteen months old.

Clouded leopards are found in dense forests. These secretive, nocturnal animals are tree dwellers, hunting monkeys, squirrels and birds, or dropping onto deer or wild boar as they pass beneath. Accomplished climbers, they are able to run down a tree head first or climb along the underside of a bough. The two to four cubs born in a litter are independent by nine months of age.

Above and opposite top right: Fishing cat, *Felis viverrina*

Opposite

Top left: Flat-headed cat, *Felis planiceps*

Bottom: Young leopard cats, *Felis bengalensis*

Small Wild Cats

Small wild cats are found all over the world, with sever species being found in Thailand, Malaysia and Indonesia These cats range in size from the rare and little known flat-headed cat, which is about the size of a domestic cat and weighs just 1.5–2 kg, to the Asian golden cat, which weighs 12–15 kg. The marbled, fishing and leopard cats have beautifully patterned skins, while the others are plain in color.

The marbled cat lives in trees and is active at night whereas the jungle cat is found only on the ground and is active during the day. The bay or Bornean red cat's distribution is limited to Borneo. The leopard and Asian golden cats are quite tolerant of people and may be found near villages. Here they make themselves unpopular by taking poultry and sheep—and goats in the case of the Asian golden cat.

The small cats mostly feed on birds and small mammals, but the fishing cat, not surprisingly, also catches fish and crabs, by crouching on a rock or ledge overhanging a river bank and scooping them out onto the river bank with a paw. Fishing cats are very aggressive and in captivity never become tame. A fishing cat once killed a leopard twice its size, and in India they are said to take small children. The marbled cat and the flat-headed cat are both rare and very shy. Little is known about the flat-headed cat, but it seems to be at home in the water and in captivity will wash its food before eating it.

The greatest threat to the small cats is the fur trade. A huge number of cat skins are required to make a single coat, and as one species becomes hard to find, the hunters catch another. This trade has devastated populations all over the world, and unless the tide of public opinion flows strongly against the use of cat fur for coats, their future looks bleak.

Above:
Asian wild dog,
Cuon alpinus

Opposite

Top:
Asiatic jackal,
Canis aureus

Bottom:
Sun bear,
Helarctos malayanus

Asiatic Jackals, Asian Wild Dogs and Bears

Asiatic jackals are slender, gray dogs which weigh 8–9 kg They are widespread in Europe, Asia and Africa, but in Southeast Asia they are restricted to Thailand. These animals are active at night, resting in holes in the ground during the day. Jackals feed on small animals, berries and fruit, although they prefer meat, including carrion, and sometimes follow tigers in order to eat what the tiger leaves behind. In Thailand jackals are usually seen alone or in pairs, but in India they form large packs. Four or five pups are born in a litter and are cared for by both parents who mate for life.

Asian wild dogs are larger, from 10–20 kg. These are found from China to Java, but are rare outside protected areas. They are active during the day and hunt big prey cooperatively in extended family packs of five to twelve animals. They will then consume large amounts of meat eating up to half their body weight in an hour. Only one female in the pack breeds, while the other adults keep the lactating female supplied with meat and help care for the young animals. The average litter size is eight.

The sun bear, found in forests from Myanmar to Sumatra, is the smallest member of the bear family weighing only about 40 kg. This bear is a good climber with strong, curved claws and hairless soles, and spend much of its time in trees. Despite their small size, sun bears are feared more than tigers, because they are very aggressive if unexpectedly disturbed. Their poor eyesight makes this easy to do. Asian black bears are about twice as heavy, weighing about 100 kg, and are found as far south as Thailand.

Both bear species are omnivorous, feeding on honey fruit, buds, insects and small vertebrates. The black bear is known to prey on animals as large as adult buffalo, while the sun bear sometimes causes serious damage to coconut plantations by eating the growing points of the palms.

Weasels, Otters, Civets and Mongooses

Opposite

Top left:
Masked civet,
Paguma larvata

Top right:
Common palm civet,
Paradoxurus hermaphroditus

Center left:
Javan mongoose,
Herpestes javanicus

Bottom:
Banded palm civet,
Hemigalus derbyanus

All weasel species, although small animals with stock limbs and short muzzles, have a reputation for grea ferocity and will tackle prey much larger than themselve They are usually nocturnal and are found in a wide rang of habitats, feeding on rats, mice, frogs, birds and domes tic fowl. The yellow-throated marten eats bees as we and is very fond of honey.

Otters are excellent swimmers and feed on fish an amphibians, but they also catch invertebrates an rodents. Some otter species favor mountain stream while others are found in lowland lakes and floode fields. Some even frequent coastal waters.

Most Southeast Asian civets are nocturnal and loo like long-nosed cats, with long tails and often a patterne coat. The binturong is the largest of the civet family, an its long, black, shaggy hair makes it look even larger. I lives in tall forest, sleeping in a hole in a tree during th day and emerging at night to feed on fruit and small an mals. Binturong have also been found swimming in rive and catching fish. The banded palm civet also forages a night, in trees and on the ground, for a range of sma prey and rests during the day in holes in tree trunk Although the feet of the otter civet are less webbed tha those of the true otter, it is a good swimmer and catche fish. Unlike a true otter, it can also climb trees.

Mongooses have long bodies, short legs and smal rounded ears, and are mostly diurnal. The Javan mor goose is probably most famous for attacking and eatin cobras. However, these form only a small part of its die with rats, birds and other reptiles being more commonl eaten. Contrary to popular opinion, the mongoose is nc immune to the snake's venom, but confuses the snak by running in circles around it, out of striking range. A the snake tires, the mongoose is able to dart in and bit its head.

Squirrels

Opposite

Top left:
Giant squirrel,
Ratufa affinis

Top right and center left:
Spotted giant flying squirrel,
Petaurista elegans

Center right:
Plantain squirrel,
Callosciurus notatus

Bottom:
Provost's squirrel,
Callosciurus prevosti

Squirrels come in all shapes and sizes and are found i most parts of the world, Southeast Asia being no excep tion. The giant squirrels are by far the largest of thi group, measuring 70–90 cm from nose to tail and weigh ing as much as a domestic cat. They are active durin the day in the high forest canopy, feeding on fruit an various leaves.

The middle layers of the forest are inhabited by th smaller arboreal squirrels, many of the genus *Callosciu rus*, which means beautiful. These animals have a hea and body length of about 20–25 cm, with tails about a long again. They feed predominantly on fruit and nuts although some species also include leaves and insects i their diet. The most striking species is undoubtedl Prevost's squirrel, with its black back and reddish bell separated by a conspicuous white stripe.

In the middle and lower layers of the forest, th smaller Sunda squirrels feed mainly on bark and sap and are often seen on tree trunks where they leave dis tinctive tooth marks. These squirrels are also found o the ground, often on fallen trees, where they feed o insects. The ground squirrels forage for fallen fruit roots and insects.

Flying squirrels are all nocturnal, and range in siz from a head and body length of about 12 cm to over 9 cm. They cannot really fly, but rather glide, stretchin out a thin, furred membrane which runs from thei front to hind legs. Using their tails as rudders, flyin squirrels glide long distances from tree to tree. They ar slower and less agile than other squirrels, because of th bulky membrane, so these squirrels may have becom nocturnal in order to avoid sharp-eyed birds of prey. Fly ing squirrels feed on fruit, leaves, insects and bark, an the species divide the canopy between them in the sam way as the diurnal squirrels, thus avoiding competin with each other for food.

Mice and Rats

Opposite

Top:
Glacier rat,
Rattus richardsoni

Bottom:
Lowland forest
Sulawesi rat,
Maxomys helwaldii

More than a quarter of the world's mammal species are rats and mice. Over 1000 species have been identified, living in almost all corners of the globe. About 200 species are found in Southeast Asia alone, in just about every available habitat.

This extremely successful animal family contains some members which are major pests and disease carriers. It is estimated that 10–20% of the grain grown in the region is consumed by rats and mice.

One common pest is the great bandicoot. This huge black and gray rat measures about 50 cm from nose to tail and is very adaptable. Bandicoots nest in burrows often in the dykes between rice fields and have frequent, large litters. The ricefield rat, a little smaller than the well-known, disease-carrying roof or house rat, is also a major pest. It is found only in association with rice fields and other plantings and is never seen in natural habitats.

Many species, however, favor natural habitats and do not pose a threat to people or their crops. The yellow rajah rat is found almost exclusively in natural habitats, living in both forests and grasslands. The coloring of this handsome rat results from the mix of short yellow fur and long black hairs. It is very common throughout Southeast Asia.

The pencil-tailed tree mouse is distinguished by its furry tail and nests inside bamboo stems, gnawing a neat round hole into the ready-made compartment. This animal also poses no threat to man.

While many species of rats and mice have very large distributions, some, like the soft-furred glacier rat of New Guinea, are known to inhabit only a few locations—in its case three wet, cold mountain sites. Nothing is known about the diet, reproduction or nesting behavior of this species.

Fruit Bats

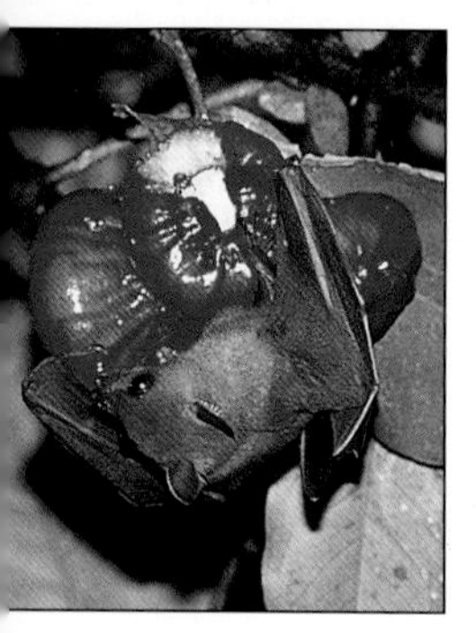

Above

Top:
Flying fox fruit bats
Pteropus vampyrus

Bottom:
Short-nosed fruit bat,
Cynopterus spinx

Opposite

Top:
Short-nosed fruit bat,
Cynopterus spinx

Bottom left:
Flying fox fruit bat,
Pteropus vampyrus

Bottom right:
Stripe-faced fruit bat,
Styloctenium wallacei

The bird-like shapes seen at dusk flitting through the air chasing insects or traveling in large flocks high in the sky are probably bats, which are widespread throughout Southeast Asia. Nearly a quarter of all mammal species are bats and around half the known bat species are found in this region.

Fruit bats come in all sizes—from the largest flying foxes, with wingspans of 2 m, weighing over 1 kg—to the smallest, with wingspans of 30 cm, weighing only 15 g. They generally have dog-shaped faces, very different from the squat faces and intricate noseleaves of the insectivorous bats.

Fruit bats have good eyesight and an excellent sense of smell. Their diet includes fruit, flowers, nectar and pollen, and they are important pollinators of many plants, including the economically important bananas, durians and kapok. Members of one group, which feed on nectar, have very elongated snouts with long, rough tongues to reach deep into flowers. In the process the bats collect pollen on their heads and shoulders which they transfer to the next flower they visit. Some plants synchronize nectar flow with bat activity—in bananas, nectar flow begins at dusk, and continues for half the night. Flowers which attract bats tend to be large, pale-colored, and open after sunset. Fruit eaters take small fruits to a nearby feeding roost where they chew and squeeze the fruit pulp in their mouths to extract the juices then spit out the dry pulp, which can be seen in piles on the ground.

Large fruit bats roost in exposed trees during the day, forming noisy colonies. Smaller species tend to roost in less conspicuous places—in holes in trees or under leaves—each species having its preferred type of roost. Rousette bats usually roost in caves and are at risk from quarrying. This may have a direct impact on durian crops in the area as their pollinators become scarce.

Insectivorous Bats

Opposite

Top:
Wrinkle-lipped bats, *Tadarida plicata*, at sunset

Bottom left:
False vampire, *Megaderma spasma*

Bottom right:
Roundleaf bat, *Hipperosideros diadema*

Bats thought of as "typical" are often insectivorous. They are small and many have intricate folds on their faces, dominated by complex noseleaves. Most insect-eating bats have poor sight, but are able to navigate and catch their prey using sound. They emit a series of high-pitched squeaks from their noses or mouths as they fly, the nose-leaves acting to direct and focus the sound. When these sound waves strike an object, they are reflected back to the bat's ears. These squeaks are too high for people to hear, but a device that lowers the frequency of sounds can transform a silent night into a veritable cacophony. This navigation system is accurate enough to enable the bats to fly in dense forest, catch an insect on the wing, or steer safely within a cave crowded with other flying bats. In response some moths have evolved the ability to issue noises of their own, on the same frequency, thus confusing the echo picture long enough to escape. Bats flying in a group must alter the frequency they use for echo location so that individuals can differentiate between their own echoes and those of the bats around them.

Insectivorous bats are an important element in the control of insect pests. In Jakarta the removal of a colony from a hospital roof space resulted in a noticeable rise in the number of patients admitted with insect-borne diseases.

Little is known about the reproductive cycle of most tropical bats. Usually a single young is born, weighing about 25% of the mother's weight. These large offspring mature quickly and are able to fly by two months of age. Since it is difficult for the mothers to fly carrying such a heavy weight, the young of some species are left together in nursery areas while the mothers forage.

The so-called insectivorous bats also include species which feed on amphibians, mice or smaller bats. Others feed on fish, flying low over streams, catching fish in their claws. These bats find their prey by homing in on the sounds they make.

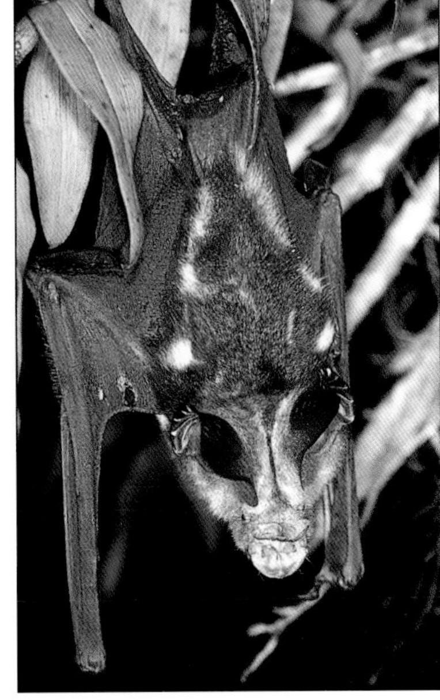

Above and opposite bottom: Common treeshrew, *Tupia glis*

Opposite top: Javan pangolin, *Manis javanica*

Treeshrews, Moonrats and Pangolins

There are eighteen species of treeshrew, all found in the forests of Southeast Asia. They are distinguished from squirrels by their longer, more pointed muzzles and more deliberate climbing. Some treeshrews bear their young in a separate nest and only visit to feed them every two days because their milk is extremely rich. The young soon grow to independence.

The common tree shrew is reddish brown. Its head and body is about 20 cm in length, with a hairy, slender tail nearly as long. Frequently seen and heard throughout the region, this shrew is very vocal. It has a loud alarm call and often makes a loud purring sound as it forages during the day for insects, spiders and worms. The pen-tailed treeshrew is named for its distinctive tail, resembling a quill pen, which is longer than the head and body and bare except for long hairs on the final quarter. This species is nocturnal, arboreal and rarely seen.

Moonrats are striking animals, with shaggy black fur on their bodies, white fur on their shoulders and face and black eye patches. They emit a pungent ammonia smell especially when frightened. They are widespread throughout the region—in forests, plantations and mangroves—but are nocturnal and rarely seen. They swim well and feed on frogs, fish, crabs and insects.

Pangolins have an unusual appearance, being covered by overlapping body scales. They are nocturnal and feed exclusively on ants and termites. It has been calculated that a single adult eats about 73 million ants a year! They lack teeth, but break apart ant and termite nests with their strong claws and use their long tongues and copious sticky saliva to draw the exposed ants into their mouths. The nose and ears can be closed and the eyes have a thick membrane as a protection against ants. If an ant should get between its scales, the pangolin kills it by grinding the scales together. The single young is carried by its mother on the base of her tail.

Above:
Black tree kangaroo,
Dendrolagus ursinus

Opposite:
Agile wallaby,
Macropus agilis

Tree Kangaroos and Agile Wallabies

Indonesia is home to many species of marsupial mammals, whose young develop in a pouch. These are mostly found on the island of New Guinea, where the fauna is closely related to that of Australia.

The agile wallaby, shown on the opposite page, is found in woodlands and grasslands, and is the largest land mammal in New Guinea, weighing up to 26 kg. It grazes on grass, also feeding on grass roots, eucalyptus leaves and figs. Agile wallabies form groups of up to ten individuals and larger aggregations are seen when food is plentiful. Several smaller species of wallaby live in the forests of New Guinea.

The tree kangaroos, genus *Dendrolagus*, have a different body shape from their relatives, and are well adapted for climbing. Their faces are bear-like and their four legs roughly the same length, with feet and hands broader and rougher than those of their earth-bound cousins. They range in weight from about 6–10 kg.

These kangaroos are accomplished climbers and have large, strong claws with which to grip branches. They can jump as far as 6 m between trees, their tails acting as rudders. In contrast, they are peculiarly awkward if they have to make their way along the ground. Tree kangaroos have varied diets; some feed mostly on fruit or leaves, others on a mixture of the two. Birds and other small animals are sometimes eaten as well.

Like all marsupials, wallabies and tree kangaroos bear very undeveloped young after a short gestation. These embryo-like offspring then climb into the mother's pouch and attach to a nipple. Most development and growth occurs here. The young remain in the pouch until they are developed enough to emerge, a process that is analogous to birth in other mammals. As illustrated in the photograph opposite, the young marsupial continues to use its mother's pouch as a refuge.

Cuscus, Gliders and Possums

Cuscus are chunky, tree-dwelling marsupials with prehensile, partly hairless tails. The island of New Guinea is home to eight species of cuscus, and Sulawesi to two very different ones. These two species are the ecological equivalents of monkeys and apes, and where they coexist, the larger usually eats more leaves and lives high in the canopy, and the smaller eats more fruit and lives lower down. Ground cuscus are partly carnivorous.

Gliders have a thin, furred membrane that stretches between their front and back limbs that can be extended to form a large gliding surface. The smaller species of glider are very maneuverable and can weave between trees before landing on a trunk up to 100 m away. Gliders are all nocturnal and fall into four dietary groups: leaf eaters, insect eaters, nectar drinkers and those that feed on sap and gum.

The sugar glider, a member of this last group, is about 30 cm from nose to tail. It feeds on wattle and eucalyptus sap, biting the bark of trees to encourage its flow. Sugar gliders also eat some pollen and insects to provide the necessary protein for growth and reproduction. These animals are common in a variety of forest types in New Guinea and often can be found resting in the tree hollows during the day.

The striped possum is one of a number of possums found in New Guinea. This squirrel-sized animal weighs less than 0.5 kg and has a striking set of black and white stripes running from nose to tail. These nocturnal animals feed on wood-boring beetle larvae, using their strong teeth to dig into the wood and their long tongues to reach the larvae. They tap tree trunks with their long claws, possibly to encourage the larvae to move so that they can be located.

Above:
Sugar glider,
Petaurus breviceps

Opposite

Top left and bottom:
Bear cuscus,
Ailurops ursinus

Top right:
Striped possum,
Dactylopsila trivirgata

Top center:
Spotted cuscus,
Spilocuscus maculatus

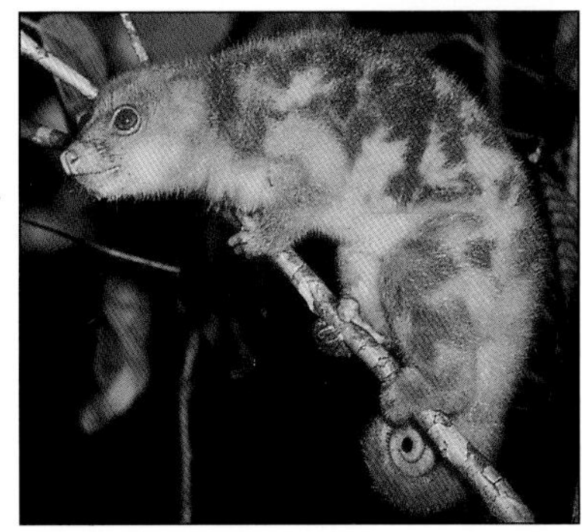

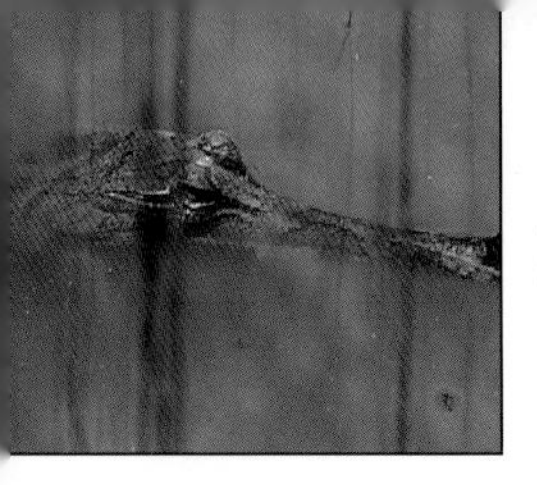

Crocodiles

Above:
False ghavial,
Tomistoma schlegelii

Opposite

Top:
New Guinea crocodile,
Crocodylus novaeguineae

Bottom:
Estuarine crocodile,
Crocodilus porosus

Crocodiles have never been popular neighbors, particu larly the sea-going estuarine crocodile, which feeds on large mammals, including people. It was reported in 1970 that a 7-m estuarine crocodile was finally killed in south ern Irian Jaya, but only after it had claimed an astounding 55 lives. In some communities crocodiles were killed whenever possible, while in others a crocodile was only killed in retaliation for taking a life. The false ghavial and the New Guinea crocodile are smaller than the estuarine crocodile and feed only on fish.

When a female crocodile is ready to lay eggs, she seeks a shady location on land where she builds a dome shaped nest of leaves or tall grass. Up to fifty or more eggs are laid in the middle of the nest, where they remain damp and are protected from direct sunlight. If the mother senses that the eggs are becoming too hot, she will cool them with a spray of urine. This temperature control is very important since the sex of the hatchling crocodiles seems to be determined by the incubation temperature. A temperature change of about 4°C makes the difference between an all-female and an all-male hatching.

Young crocodiles are carried gently in their parents mouths to a secluded nursery area where they stay for a month or two, catching large insects, fish and frogs. Dur ing this time the parents guard the hatchlings closely and are very aggressive.

Trade in wild crocodile skins is centuries old and they are still a valuable commodity. Recent hunting has taken a severe toll on wild stocks, consequently crocodiles are now absent from many areas or much reduced in num bers. In an effort to reduce the threat to wild populations a few crocodile farms have been set up. Wild-caught young animals are brought to these farms and skinned when they reach an economic size.

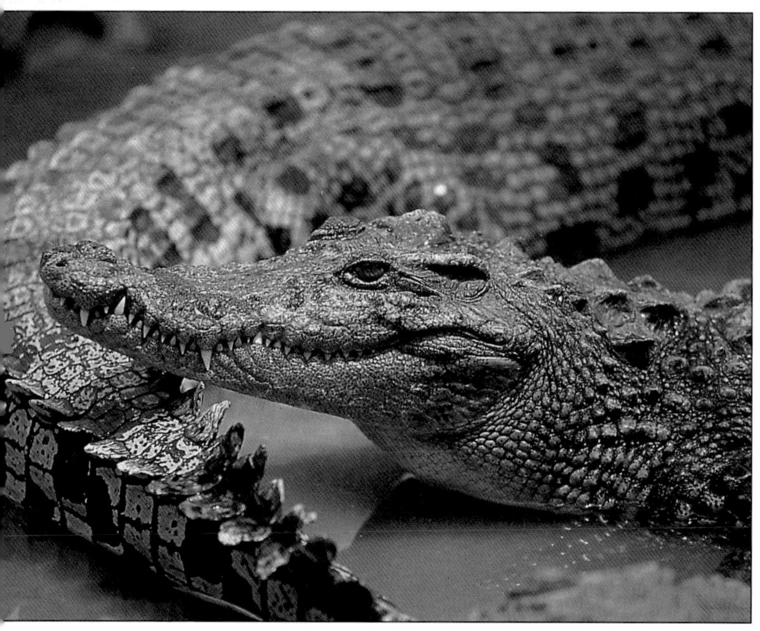

Above

Top:
Tokay gecko,
Gekko gecko

Bottom:
Sun skink,
Mabuya multifasciata

Opposite

Top:
Common flying lizard,
Draco volans

Bottom:
Blue-tailed skink,
Emoia cyanura

Geckos, Skinks and Lizards

Lizards are found throughout the region except on higl mountain tops. Those most commonly seen are the sev eral species of gecko which frequent houses and provide valuable service by eating flying insects. In the forest shiny skinks bask conspicuously in the sun, flying lizard glide from tree to tree, and occasionally a large monito lizard may be heard crashing through the undergrowtl and splashing into a river.

Geckos are familiar to anyone who has spent time ir the region. The feet of these small lizards are covere with flaps of overlapping skin, which are covered witl minute, closely-set, hooked hairs. These make contac with surface irregularities, rather like Velcro, and enabl the gecko to cling to walls and ceilings. Geckos often tak up residence behind picture frames on a wall, emerging a night to feed off insects attracted to the light. A numbe of species may be present, sharing the available resource by catching prey of different sizes and adopting differen hunting techniques. Some species lie in wait, while other actively search for prey. The larger tokay gecko is als associated with human habitation, and feeds on insect and small vertebrates, including smaller geckos.

Skinks have short legs, are often brightly colored, an have a shiny appearance because of their very small bod scales. Some species favor forests, while others may b found in gardens and agricultural areas. In either habita they feed on small invertebrates.

Another common lizard is the flying lizard, frequentl seen in gardens and agricultural areas. The ribs of thes lizards project from the body to support two flaps of ski which can be extended at will. These lizards leap fror great heights and make controlled glides of considerabl distances. On landing, their cryptic coloration render them almost invisible. Males display conspicuously o tree trunks by extending their bright yellow throat flap and doing pushups to impress the females.

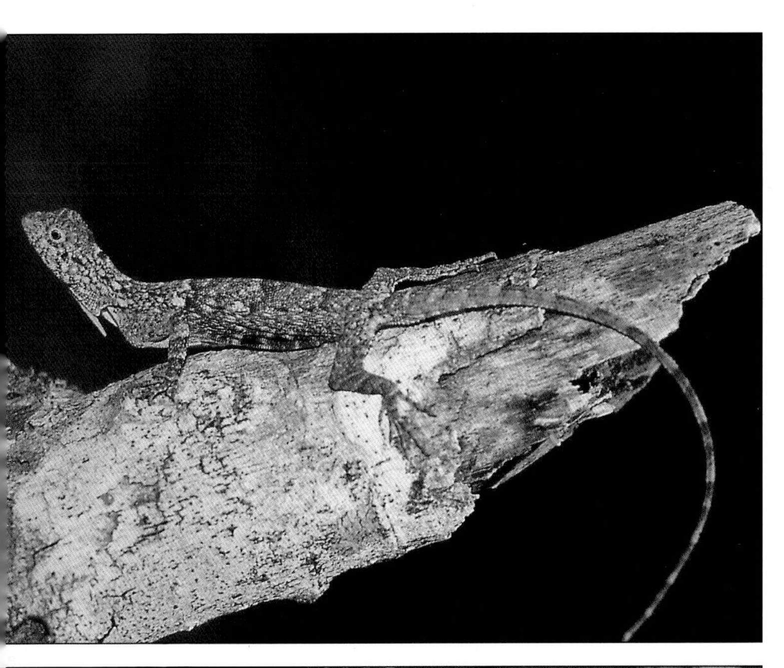

Above

Top:
Komodo dragon,
Varanus komodoensis

Bottom:
Monitor lizard,
Varanus salvator

Opposite

Top:
Komodo dragon,
Varanus komodoensis

Bottom:
Sail-fin lizard,
Hydrosaurus amboinensis

Monitor Lizards

The most famous lizard in the region is probably the Komodo dragon, which reaches nearly 3 m in length and weighs up to 50 kg. These huge lizards are restricted to Komodo and neighboring small islands in central Indonesia. Adult dragons occupy burrows in dry creeks to escape from the fierce heat of the day and the cool of the night; young animals also use holes in trees. Komodo dragons are most active during the day, when they hunt live prey or search for carrion using their keen sense of smell. A hungry Komodo dragon can eat up to 80% of its own body weight in food in a single day. Their prey is predominantly deer and pigs, but many other animals are also taken, and occasionally humans have been attacked. These lizards can cover 100 m in 20 seconds, making them fearsome hunters. The females lay about eighteen eggs over a period of days during the dry season. The 30-cm hatchlings, with their speckled camouflage, are mainly arboreal predators feeding on insects and other small lizards. At about a year old they graduate to eating birds and rats.

Monitor lizards, relatives of the Komodo dragon, are found throughout the region. They may grow to 2.5 m, but are much slighter in build than the dragon and are not considered dangerous to man, preferring to flee rather than fight. The water monitor is an accomplished climber and swimmer, often found near water. It feeds on small animals, taking chickens and ducks from villages as well as catching wild food.

Another spectacular species is the sail-fin lizard of eastern Indonesia, which reaches just over 1 m in length, only one-third of which is its body. This shy animal is most frequently seen basking on a branch overhanging water, with its feet dangling down on either side. The "sail" is held erect by projections from the backbone. This is the only lizard that eats leaves, but it also feeds on insects and fruit.

Snakes I

Above

Top:
Common blind snake, *Ramphotyphlops braminus*

Bottom:
Python hatchlings

Opposite

Top left:
Paradise tree snake, *Chrysopelea paradisi*

Top right:
Young green tree python, *Chondropython viridis*

Center right:
Reticulated python, *Python reticulatus*

Bottom:
Adult green tree python, *Chondropython viridis*

Snakes generally receive a bad press, but there are many beautiful snakes in the region that are harmless to humans. All snakes are carnivorous; a few species live on worms, snails or insect larva, but most feed on larger prey and must dislocate their jaws in order to swallow them. Many snakes swallow their prey alive, but the venomous snakes kill or paralyze it first and the constrictors squeeze their prey until it cannot breathe. Some snakes will eat anything of an appropriate size, but most have clear food preferences, and many species are important in reducing rodent pests in villages and fields.

Blind snakes are small, worm-like creatures which may easily be mistaken for earthworms as they burrow in search of small invertebrates.

Pythons are well-known constrictor snakes. Their protective coloring camouflages them, making it difficult for prey to see them approaching. These snakes have heat-detecting pits on either side of their heads that enable them to seek out warm-blooded animals in the dark. Large pythons may reach nearly 10 m, but these huge old animals have mostly been shot for their skins, and may be a thing of the past. Pythons lay 10–100 eggs and are unusual in that they will guard their nests. No snake takes care of its hatchlings.

Tree snakes are particularly beautiful. They are back-fanged, with small, grooved poison fangs at the back of the upper jaw, but the venom is not poisonous to people. These snakes are very active and able to climb vertical trunks with ease, or to leap from one tree to another by coiling like a spring and then straightening suddenly. These snakes feed mostly on lizards, sometimes battling for up to an hour with a large forest gecko before their mild venom subdues the victim. Some tree snakes are also known as flying snakes. They can flatten their bodies into a concave shape and trap a pocket of air under their bellies as they fall, to help them glide for long distances.

Snakes II

Above:
Black spitting cobra,
Naja sumatrana

Opposite

Top left:
Thai cobras,
Naja kaouthia

Top right:
Banded sea snake,
Leioselasma cyanocincta

Center right:
Blue coral snake,
Maticora bivirgata

Bottom:
Shore pit viper,
Trimeresurus purpureomaculatus

Vipers are all venomous and have long fangs. Their broad triangular heads, distinct necks and rather thick bodies can be easily recognized. The pit vipers have a heat-sensitive pit on each side of the head which they use to locate warm-blooded animals. Most vipers bear live young.

Sea snakes are all venomous and most never leave the water. Their tails are flattened from side to side like paddles, making them easy to distinguish. They feed on fish and in turn are preyed upon by sharks, eels and white-bellied sea eagles. Most sea snakes bear live young, as eggs laid in the sea would be subject to heavy predation. Sea snakes are not particularly aggressive, but will bite if stepped on or disturbed in any way.

King cobras or hamadryads are the largest venomous snakes in the world, reaching over 5 m in length. Cobras give a warning display, rearing their heads up and spreading their hoods, hissing, when threatened. This snake makes a loose nest of leaves, with a lower compartment containing the eggs and an upper one that is inhabited by the snake. The compartments are separated by a layer of leaves. The eggs are guarded by the female, but even nests in well-populated areas may go undetected, so their reputation as aggressive animals seems to be largely unfounded. However, their bite is dangerous. They feed on other snakes and sometimes monitor lizards.

Coral snakes are small, slender and brightly colored. When threatened they will raise their tails and writhe about, displaying their colorful bellies as a warning. Some species have extremely long venom glands and are able to inject a considerable amount of venom. Despite this they have very small mouths and so are not usually considered dangerous. Coral snakes lay eggs and feed on other snakes.

Marine Turtles

Above and opposite bottom: Green turtle, *Chelonia mydas*

Opposite top: Hawksbill turtle, *Eretmochelys imbricata*

The leatherback, the largest of the marine turtles, can weigh up to a ton and has a dark-brown, ridged carapace up to 2.5 m long. The most common species is the green turtle, which grows to about a meter in length. About the same size, the loggerhead turtle is brown in color, with a massive head. The smallest is the olive Ridley turtle, with a shell length under 70 cm. The hawksbill turtle is just a little larger than the olive Ridley and has a thick shell that is used commercially for "tortoise shell."

Turtles all look superficially similar yet they feed on a wide range of foods. The enormous leatherbacks live on a diet consisting solely of jellyfish and this surprising food source sustains them on their lengthy migrations across the oceans, during which they swim thousands of kilometers. They are also known to eat floating plastic bags, a fatal mistake. The loggerhead turtle prefers crustaceans and mollusks; the olive Ridley turtle feeds on crabs and shrimps in shallow seas; the hawksbill's diet includes a range of invertebrates associated with coral reefs; and the green turtle is a vegetarian which feeds on beds of sea grass in shallow coastal waters.

All marine turtles are thought to return to the beaches where they hatched to lay their eggs. The females then return to the sea, and the hatchling turtles have to run the gauntlet of a variety of predators on shore (such as seabirds and dogs) as they try to reach the sea, and then face predatory fish offshore. People collect turtle eggs throughout the region as food, and it is feared that insufficient turtle numbers are left to maintain the populations, since mortality is very high for baby turtles. Also many adult turtles are killed for food and are inadvertently caught in fishing nets where they get tangled and drown.

Marine turtles are protected by international treaty. Turtle crafts and curios, although widely available in some countries, may not be taken across national borders.

Above

Top:
Giant tree frog,
Litoria infrafrenata

Bottom:
Tree frog

Opposite

Top:
Gliding frog,
Rhacophorus nigropalmatus

Bottom:
Asian horned toad,
Megophrys monticola

Frogs and Toads

From tiny tree dwellers 0.25 cm long to giants of ove 20 cm, frogs and toads are a very diverse group. They ar well-represented in Southeast Asia, where there are ove 500 different species. Tropical rainforest offers the greatest diversity, providing numerous niches with the higl humidity and standing water that many species need fo breeding. A few species, like the crab-eating frog, occu in coastal swamps; others live in mountainous forests u to 2000 m above sea level.

Tree frogs have sucker-like discs on their toes, whicl enable them to cling to flowers, branches and leave when climbing. Some lay their eggs in foam nests amon twigs or leaves that hang over water. Heavy rain disintegrates the nest and washes the tadpoles into the wate beneath. Other frogs, such as the black-spotted stick frog, named for the copious sticky substance it secrete as a deterrent to predators, live exclusively in the leaf litter. This species lays its eggs in the cups of pitcher plant on the ground. The tadpoles do not feed, but live of their yolk sacs until they have turned into frogs. Th horned toad is also a litter dweller, and is well camouflaged by its leaf-like appearance.

The crab-eating frog lives in urban and agricultura areas as well as mangrove forests. It feeds at night or small animals, including crabs, and is tolerant of sal water. Another unusual species is the gliding frog whicl has webbing between its toes. When it leaps, it spread this webbing, thus gliding from tree to tree to escap from predators.

Many frogs and toads are quite difficult to identify because the members of a species are very variable ir coloring and even some aspects of body shape. In addition many are able to change their skin coloration tc blend in with their surroundings. Their calls, however, are all distinctive, and add a melodious element to the evening chorus.

Index